McCabe Renewal Center
2125 Abbotsford Ave.
Duluth, MN 55803-2219

Caves
and
Canyons

This book is dedicated
to my sister Ava,
and to all those who
have trodden within
the Grand Canyon . . .

Caves
and
Canyons

*a refreshing journey of the spirit
to inner realities*

Illustrated by De Grazia
Words by Sister M. Angela Toigo, O.S.B.

Published by
Benedictine Sisters
3888 Paducah Drive
San Diego, California 92117

Benedictine Sisters of Perpetual Adoration
San Diego, CA 92117

Copyright © 1979
All Rights Reserved

Manufactured in the U.S.A.

Illustrations created especially for this book, and
are reproduced with special permission of the
De Grazia Gallery In The Sun,
6300 North Swan, Tucson, Arizona 85718

First Printing, 1979

Library of Congress
Catalog Card Number 79-54772

ISBN 0-913180-02-5

*The Lord said to Moses:
"Behold there is a place by me
where you shall stand upon the
rock; and while my glory passes by
I will cover you with my hand
until I have passed by; then I
will take away my hand, and you
shall see my back; but my face
shall not be seen."*

Exodus 33: 21-23

"...and the Rock was Christ."

1 Corinthians 10:4

Several poems in this book were printed in SPIRIT & LIFE Magazine, published by the Benedictine Sisters of Perpetual Adoration, Clyde, Missouri 64432.

"no mad"

hewn and hallowed-out
to bear the howling wind and wilderness
   upon bare-breasted tenderness.

such Unrelenting Desert-Love
pursuing all the wasteland
where heart-shod of idols
   I roam with God
   amid the caves and canyons
   of my soul . . .

along the Sacred Path

at the foot of the mountain
    I saw a face set as flint
    against the sun.

it was Courage . . . waiting for me.
    "Take hold of Me and pursue your dream –
    that haunting inner vision.

Be not guilty of numbing your dream.

You think your dream is for tomorrow,
but I say, with Me you will find
              your dream this day!

though you burn me
You are Great Tenderness . . .

of heights and depths

elusive vision of deity
ravaging the heart of me.

beyond the realms of sky and sea,
can I hold Thee in my frailty?

There are days when I am called
to dip the spirit of my being
    into a chalice
    of chiseled rocks
strewn in circumference.

to be still
and feel
    the throb
    of mother earth
    rub against me
and claim me as her clay.

rays of light
play upon the stones
    until fingered sunshine
    touches my heart
and then
I see myself
    'cornered' in Your Light
              alone with You.

Do you feel the breath of my desire
    powerful enough
    to break
the impenetrable rock of Your mystery
of which I am a part?

and thus,
a solitary sat and stared at a rock . . .

ancestors

cry out your tears . . .
    the dreams of my people
    lie wind-swept
    at earth's edge.

let your heart raise up
the burnt-out hopes,
for I hear Thunder howling
through the hills
and have already seen
the Shadow of the Great Spirit
hallowing the land.

    for we are the mountains,
    and we are the fields,
    and we are the running streams . . .

a hermit in the meadow

There are days
    when I am called
    to rest my being
    amid
    shades of green.

blades of grass
    myriads upon myriads —
    touch me
    and I feel
    my heart
        fingered
        by Your Presence.

deep in your arms
    the stillness searches
    inner mansions
    of divinity.

light and shadows
    peacefully play
upon the breast of mother earth

while
    I weep
for those who do not understand
God's temple of green . . .

O child within the soul of me,
O burden thou breakest the heart of me.
   none but One
   can thy panting hear,
   wandering,
   crying for the breast of God.

raindrops

Divinely touched
myriads of melted crystals
   still warm
from heaven's hearth
streak the sky
    in processions of wonder.

   crisp
splashing sparkles of eternity
come tenderly as tears fall
upon the window of my heart.

today,
   I have walked
among the raindrops . . .

Gather into the fields of Your heart
all the moments when we ache with love,

loosen the fears that make us run
from what we want,

hear the cries of our secret burdens.

O spare us,
    from living on the edge of reality.

O come,
    Great Harvester of dreams . . .

a man named Enoch

my heart is borne
    among the stars
    laden with the depth
    of all that is
    into
    the realms of what will be.
dream deep, my soul,
                      dream deep.

You can reject my visions
    but you will never
    run me out of dreams.

before the beginning
beyond forever,
    an elusive 'now'
    in measured time

becomes you
    and me
    and all the flesh
    and all the grass
    and all the rock

journeying into realms
of infinite deity.

Jubal

Genesis 4: 21

O father of all who play
the lyre and the flute . . .

You are descendant of Cain
and son of Lamech . . . a family
branded by bloodshed, forehead marks,
and forgiveness.

> *soothe the spirit of my people, Jubal.*
> *I place tenderness and thunder in your fingers*
> *Lay bare in melodies*
>   *the rhythm of a people's soul!*

When you live amid rocks
the heart is bound to cry out
   for flowers . . .

touched
by early evening stars
    I prostrate
before the God of Twilight,

the sunset is burning incense . . .

leaves
once aflame
in autumn sun-dance
    lay burnt-out.
        Color-all spent,
(and God gathering
          the fragments . . .)

snowflakes

how beautiful,
how beautiful . . .

I can feel
   the smile
of mother earth
as she is adorned
   with lace
woven in looms
from frozen flashes
   of light.

dazzling diamonds
crystal cold,
    divinely
    fingered and fashioned,

    sparkle and sing,
    and gracefully glitter
as they
    bounce and bump
    and joyfully jump

into the open arms
of awe-struck children!

how beautiful,
how beautiful . . .

deep, dark nights
    hallowing the thoughts of sages,
wind pushing with primal force
    that I may see
    the breath of me
    in coldness,
and tread among the other footprints
journeying to Spring.

O God,
    the soul needs a bit of winter.

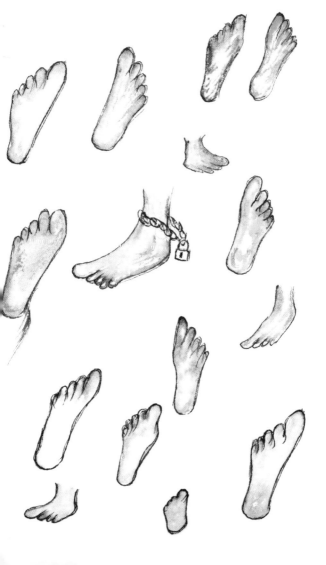

Beware when Sky-Wind comes
  shaking the mountain.

God-Breath will blow
  our illusions away . . .

howling wind of primal force
now Fire-Breath
within the soul of me,
burning out images
that bind the heart in fear
to risk the awesome emptying
of knowing God in God.

O set me free
unto such wildness . . .

Q

Walker Lithocraft • Tucson, Arizona

McCABE RENEWAL CENTER
2125 Abbotsford Ave.
Duluth, MN 55803